Take Off with

TIME

About This Book

The activities, puzzles, and games in this book about time have been designed for an adult and child to enjoy together. Take time to find out the many opportunities they provide for learning about time and how we measure it.

Each page deals with a topic that children will be introduced to in the early years at school. The pictures are of familiar objects and everyday situations that will help children to realize that time is not just about clocks, but that it is an important part of life. Children learn most effectively by joining in, talking, asking questions, and solving problems, so encourage them to talk about what they are doing and to find ways of solving the problems by themselves.

The games and activities in the "Take Off" loops will give children a chance to practice and develop the new skills they have been introduced to on that page. You may find it helpful to complete the "Take Off" activity, **Make a clock** (page 29), before you look at the pages that deal with telling time.

Use the book as a starting point. Look for other occasions to learn about time; for example, point out clocks when you are at stores, see how long daily tasks take, and note the times of favorite television programs. Make sure that it is not only easy to take off with time but also fun!

Published by Raintree Steck-Vaughn Publishers, an imprint of Steck-Vaughn Company

Library of Congress Cataloging-in-Publication Data

Hewitt, Sally.
    Time / Sally Hewitt.
        p.   cm. — (Take off with)
    ISBN 0-8172-4111-6
    1. Time — Juvenile literature.   2. Time measurements — Juvenile literature.
[1. Time.   2. Time measurements.   3. Clocks and watches.]   I. Title.   II. Series:
Hewitt, Sally. Take off with.
QB209.5.H49   1996
529'.7—dc20   95-18464
                                                                          CIP
                                                                          AC

Printed in Hong Kong
Bound in the United States
1 2 3 4 5 6 7 8 9 0 LB 99 98 97 96 95

Take Off with

TIME

Sally Hewitt

RSVP

**RAINTREE
STECK-VAUGHN**
PUBLISHERS
The Steck-Vaughn Company

Austin, Texas

Acknowledgments

Editorial: Rachel Cooke, Kathy DeVico
Design: Ann Samuel, Joyce Spicer
Production: Jenny Mulvanny, Scott Melcer
Photography: Michael Stannard
Consultant: Peter Patilla, formerly Senior Lecturer in mathematics education, Sheffield Hallam University
Artwork: Clinton Banbury Associates and Kathy Baxendale (pages 14 and 29)

The author and publishers would like to thank the following companies for their help with the objects photographed for this book:
John Lewis Partnership, pages 8, 9, 11, 12, 13, 15, 16, 18, 20, 21, 26, 27 and 28; NES Arnold Limited, pages 9, 24 and 25; Tridias, 6 Bennett Street, Bath, BA1 2QP, 0225 314730, pages 9, 10, 11, 17, 18, 19, 22 and 27; Which Watch, pages 19 and 28.

For permission to reproduce copyright material, the author and publishers gratefully acknowledge the following:
Page 10: (mouse and fox) Hans Reinhard/Bruce Coleman Limited, (moth) Andrian Davies/ Bruce Coleman Limited, (bat) Frank Greenaway/Bruce Coleman Limited. **Page 16:** (bottom left) Robert Harding Picture Library, (bottom right) Robert Bosch Domestic Appliances Limited. **Page 16:** (bottom left) © John Coletti/Stock Boston **Page 17:** (top left and bottom right) Robert Harding Picture Library, (center left) Sony, (top right) © Tony Freeman/PhotoEdit. **Page 19:** (top right) Robert Bosch Domestic Appliances Limited. **Page 25:** Robert Harding Picture Library.

Contents

Daytime

Daytime begins when the sun rises in the morning.
It ends when the sun sets in the evening.
Look at the pictures to find out what Joey does in a day.

Morning is the first part of the day.
What does Joey do in the morning?

Noon is the middle of the day.
Joey has his lunch at noon. What else does he do?

Afternoon comes between lunchtime and evening.
What does Joey do after
painting his picture?

Evening is the end of the day.
What is the last thing Joey does in the evening?

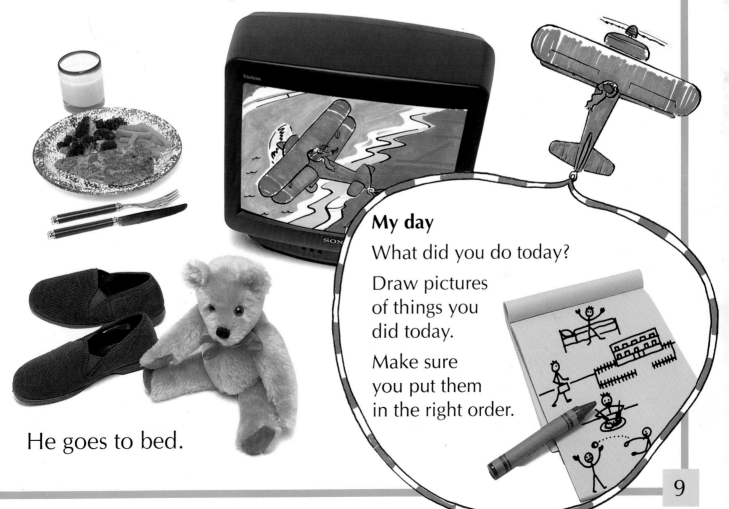

My day

What did you do today?

Draw pictures
of things you
did today.

Make sure
you put them
in the right order.

He goes to bed.

Nighttime

Nighttime begins when the sun sets,
and it gets dark.
Sometimes you can see the moon
and stars.
Nighttime ends when the sun
rises again in the morning.

At night, you go to
sleep after a busy day.
But not everyone
is sleeping.

Some animals look for food at night.

Some people work all night.

Who works at night to get these things ready for the morning?

Days of the Week

There are seven days in a week.
Do you know the names of the days of the week?

These pictures show you what Joey is doing this week.

swimming

going to
school

going to
the park

going to a party

visiting a friend

Monday	 
Tuesday	 
Wednesday	 

Thursday			
Friday			
Saturday			
Sunday			

Saturday and Sunday
are called the weekend.
Can you guess why?

The day today

What day is it today?
What is Joey doing today?
What did he do yesterday?
What is he going to do tomorrow?
Pretend another day is today,
and ask the same questions.

Ask the same questions
about yourself.

A Year

In a year, there are
365 days,
52 weeks, and
12 months.

Every four years, a year is
a day longer.
It has 366 days.
We call this a leap year.

This year circle shows the names of the 12 months
and how long they are. It also shows the four seasons.

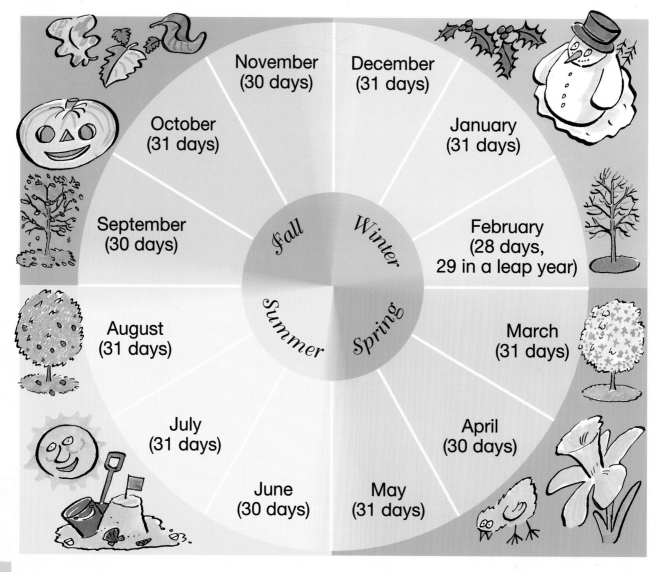

November
(30 days)

December
(31 days)

October
(31 days)

January
(31 days)

September
(30 days)

Fall

Winter

February
(28 days,
29 in a leap year)

August
(31 days)

Summer

Spring

March
(31 days)

July
(31 days)

April
(30 days)

June
(30 days)

May
(31 days)

A whole year passes
between birthdays.
Joey's birthday is in January.
This year he will be six.

Is his birthday in the spring, summer, fall, or winter?

Which month comes just before his birthday?
Which month comes right after his birthday?
How old will Joey be next year?

Ask the same questions about your birthday.

A day to remember

On your birthday, make a form like this
one, and fill it in.

My birthday is on _____ .
I am _____ years old.
I am _____ inches (cm) tall.
I weigh _____ pounds (kg).
My shoe size is _____ .

Do the same next year, and see
how much you have grown.

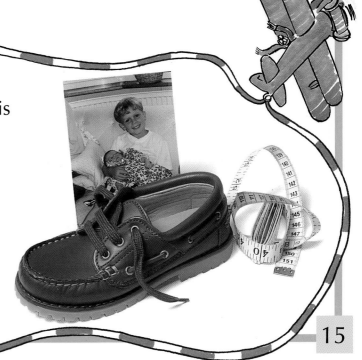

Clocks Everywhere

We often need to know how long it takes to do something or what time of day it is.
We measure time with clocks.

Look for clocks, and you will see them everywhere you go.

An alarm clock wakes you up in the morning.
What other helpful jobs do these clocks do?

alarm clock

wristwatch

town hall clock

oven clock

children's clock

Arrivals

airport clock

railroad station clock

clock on a video recorder

A clock measures time
in hours, minutes,
and seconds.

Watching the clock

You could be late for school,
if you didn't know the time.

How many other things in
your day could you miss
if you didn't know the time?

Seconds

A second is a very small amount of time.
It takes about one second to clap once or
to stamp your foot.

Here are some things to do that take about one second.
Try them, and see.

one bang on
a drum

one bounce of
a ball

one sip of water

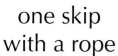

one skip
with a rope

What else can you do that
takes about one second?

These timers measure seconds ticking by.

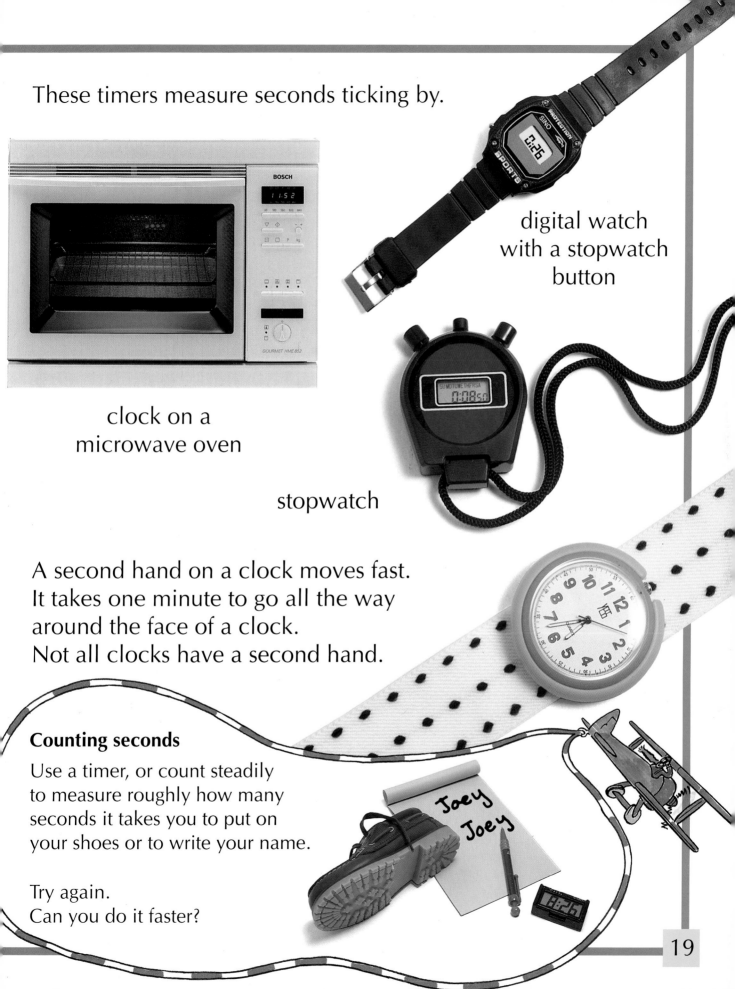

digital watch
with a stopwatch
button

clock on a
microwave oven

stopwatch

A second hand on a clock moves fast.
It takes one minute to go all the way
around the face of a clock.
Not all clocks have a second hand.

Counting seconds

Use a timer, or count steadily
to measure roughly how many
seconds it takes you to put on
your shoes or to write your name.

Try again.
Can you do it faster?

Minutes and Hours

There are 60 seconds in one minute.

This is the minute hand on a clock. It goes too slowly to see it move.

It takes one minute to move this far around the clock.

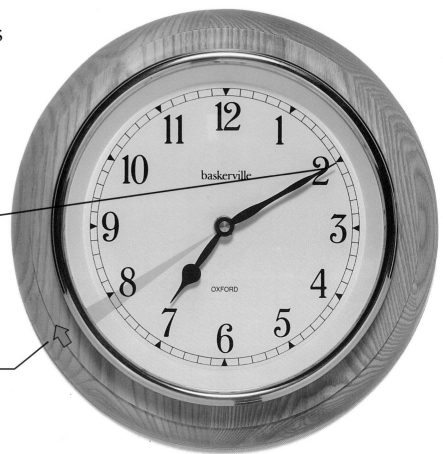

These are the minutes on a digital clock.

You can see them change every time a minute passes.

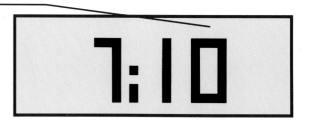

It takes one minute to take your temperature.
It takes ten minutes to hard-boil an egg.

There are 60 minutes in one hour.
It takes the minute hand one hour
to move all the way around
the clock.

This is the hour hand
on a clock.

It moves even more slowly
than the minute hand.
It takes one hour to move from
one number to the next.

This number tells you
the hour on a digital clock.

It takes the small hand 12 hours
to go all the way around the clock.

There are 24 hours in a day.
So it goes around twice each day.

The small hand points at the 12
at midnight and at noon.

An hour is a long time.
Notice what you are doing during
the day as the hours pass.

Telling the Time

The position of the hands on a clock's face tells you the time.

When the minute hand is pointing to the 12, it is a certain hour exactly.

The hour hand is pointing to the 4.

Both hands together tell you that it is four o'clock.

The numbers on a digital clock tell you the time.

When the minute number shows 00, it is a certain hour exactly.

The hour number is 4.

Both numbers together tell you that it is four o'clock.

Try telling the time on these clocks.
Do any of them show the same time?

Clock hands always follow the numbers in order around the clock. We call the direction they move in, clockwise.

We often divide hours up into halves and quarters.

The minute hand has moved a quarter of the way around the clock's face. The hour hand is just past the 4. We say the time is now quarter after four.

The minute hand has moved halfway around the clock. The hour hand is halfway between the 4 and the 5. We say it is half past four.

The minute hand has moved three-quarters of the way around the clock. It has a quarter still to go. The hour hand is almost on the 5. We say it is quarter to five.

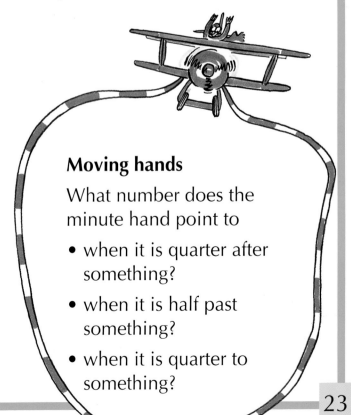

Moving hands

What number does the minute hand point to

- when it is quarter after something?

- when it is half past something?

- when it is quarter to something?

23

Passing Minutes

It is four o'clock.

One minute goes by.

The minute hand has moved one minute past four o'clock.

It takes the minute hand five minutes to move between each number on the clock's face.

Put your finger on the blue 5. Move it clockwise around the clock, counting the minutes in fives.

The time on this clock is five minutes after four o'clock. We say it is five after four.

4:05

What is the time on this clock? Five after four. It is the same. Were you right?

Tell the time as the minute hand moves around the clock's face.

The digital clock is showing the same time.

ten after four
(4:10)

twenty-five after
four or
four twenty-five
(4:25)

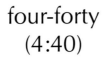

four-forty
(4:40)

four fifty-five
(4:55)

Before clocks were invented, people used other things to measure time.

Hands and numbers

Point to the place the minute hand would be on a clock's face when a digital clock shows these times:

4:15   4:30   4:45

It takes one hour for the shadow to move from one number to the next on this sundial.

It takes one minute for the sand to pour through this timer.

25

How Long Does It Take?

Joey started to make his sandwich at three o'clock.

He finished making it at five after three.

The minute hand moved five minutes around the clock while he was making his sandwich.
He took five minutes.

How long did Joey take to do each of these things?

Joey took
- 20 minutes to do his painting.
- 30 minutes to build his castle.
- 15 minutes to finish his puzzle.

Were you right?

Passing time

Next time you brush your teeth, get dressed, or eat your breakfast, look at the clock to see how long you take.

Find out how long you take to do some other things.

What Time Is It?

The time can be written as words or as numbers.
Can you read the times that are written on this page?
Follow the string from each word to see if you are right.

quarter after eleven

six-forty

twenty after seven

five o'clock

ten-thirty

ten after eight

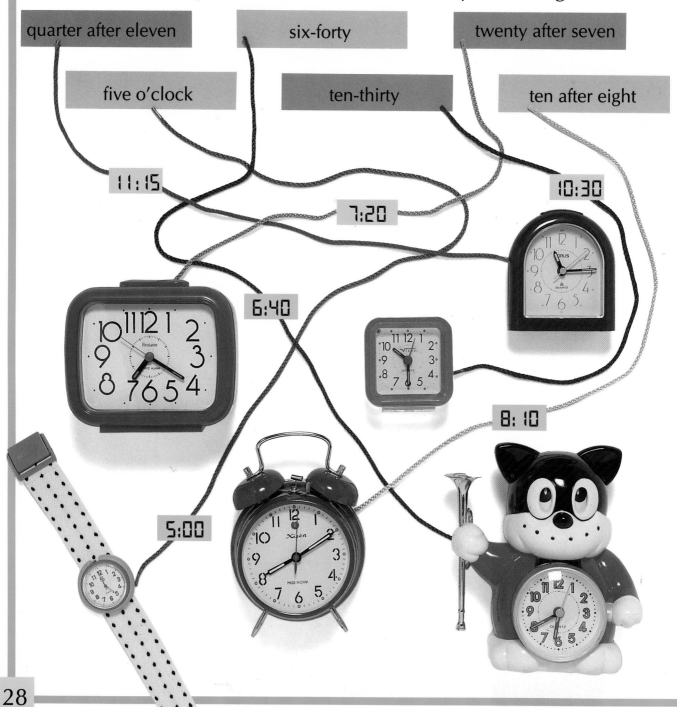

11:15

7:20

10:30

6:40

8:10

5:00

Make a clock

Use the clock face and hands on this page to make a clock. It will help you practice telling the time.

You will need:
- a piece of paper
- a pencil
- scissors
- a black felt-tip pen
- a paper fastener

12 11 1 10 2 9 3 8 4 7 6 5

1. With a pencil, trace the clock face, hands, and minute markings onto the piece of paper.

2. Cut them out.

3. Draw over the minute markings with the felt-tip pen.

4. Now write in the numbers. Be careful to put them in the right place.

5. Attach the hands to the middle of the clock's face with a paper fastener.

6. Move the hands, and tell the time!

Birthday Time

Here is a January calendar.
Joey's birthday is in January.
Read the questions below.
Use the calendar to answer the questions.

Sunday	Monday	Tuesday	Wednesday	Thursday	Friday	Saturday
	1	2	3	4	5	6
7	8	9	10	11	12	13
14	15	16	17	18	19	20
21	22	23	24	25	26	27
28	29	30	31			

1. Joey's birthday is January 22.
   What day is his birthday?

2. Joey will have a birthday party on the third Saturday.
   What is the date of his party?

3. Joey's parents gave him an early birthday present.
   He will take piano lessons every Monday.
   How many lessons will Joey have in January?

4. Joey's grandmother lives in another state.
   She wants to come to Joey's party.
   She will arrive at the airport on January 18.
   What day will she arrive?

5. Joey's grandmother will stay one week.
   What date will she leave on?

Joey is getting ready for his party.
Read what time he does everything.
Can you find the matching clock?

6. Joey decorates the house at 10:50.   a.

7. Joey serves birthday cake at 3:15.   b.

Look at the clocks.
Read what Joey does to get ready for his party.
What time will the clock say when Joey stops?

8. 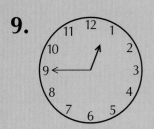 Joey blows up balloons at 11:20.
It takes him 20 minutes.
What time will the clock
say when he stops?

9. 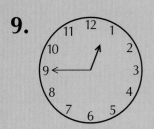 Joey walks to the store to
get the cake at 12:45.
He is gone for 30 minutes.
What time will the clock say
when he gets back home?

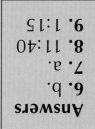

Answers
6. b.
7. a.
8. 11:40
9. 1:15